Alfred Wegener Uncovering Plate Tectonics

Greg Young, M.S. Ed.

Earth and Space Science Readers:
Alfred Wegener: Uncovering Plate Tectonics

Publishing Credits

Editorial Director
Dona Herweck Rice

Associate Editor
Joshua BishopRoby

Editor-in-Chief
Sharon Coan, M.S.Ed.

Creative Director
Lee Aucoin

Illustration Manager
Timothy J. Bradley

Publisher
Rachelle Cracchiolo, M.S.Ed.

Science Contributor
Sally Ride Science

Science Consultants
Nancy McKeown,
Planetary Geologist
William B. Rice
Engineering Geologist

Teacher Created Materials
5301 Oceanus Drive
Huntington Beach, CA 92649
http://www.tcmpub.com

ISBN 978-0-7439-0560-2

Reprinted 2012

Table of Contents

Early Life 4

Becoming a Scientist..6

Puzzle Pieces...12

Proving the Theory..16

Wegener's Theory Now..22

Geophysicist: Mary Lou Zoback 26

Appendices..28

Lab: An Eggsample Look Inside Earth 28

Glossary...30

Index 31

Sally Ride Science...32

Image Credits 32

Early Life

Alfred Wegener was born in Berlin, Germany, in 1880. His father, Richard, was a minister who ran an orphanage. His mother, Anna, took care of the home and family.

When he was a boy, Wegener hoped to explore Greenland one day. He hiked, skated, and walked to build his strength. He wanted to be strong enough to explore! When he walked, he pretended he was on a great **expedition**. He always knew that one day he would be an explorer.

When he grew up, he studied **astronomy**. He even got his Ph.D. in this field. That is the highest degree a person can earn.

Did You Know?

Greenland is the world's largest island. Most of it is covered with ice. Sailors from Iceland discovered it in A.D. 982. People, called the Inuit, had lived there centuries before.

Greenland is beautiful, but it is challenging to explore.

He found, however, that astronomy was not earthy enough for him. He still wanted to explore. He wanted to know more about the earth. So he began to study **meteorology**. That is the science of weather. This brought him closer to two other interests. He wanted to explore Greenland. And he wanted to experiment with hot air balloon flight.

Becoming a Scientist

Wegener had a brother named Kurt. They experimented with kites and hot air balloons. The two brothers were in a balloon contest in 1906. They set a world record. They flew for 52 hours!

That same year, Alfred made his first trip to Greenland. He was invited because of his balloon flight. He went as the official meteorologist for the trip. He used kites and balloons to study the weather. Meteorology seemed to be his calling.

Wegener had another dream. He wanted to teach at a college. This dream also came true. A university in Germany hired Wegener. He taught meteorology there.

Wegener became well known for his work. He was well liked by the students and other teachers. They thought he was full of great ideas.

It was here that he came up with his most famous idea. He called it **continental drift**. It led to the study of **plate tectonics.** It is a **theory** about how the land and seas on Earth are formed and change. Wegener's big work in this area was to come later.

Did You Know?

Hot air balloons don't have engines or motors. They don't need electricity. They work because of basic science. Warmer air rises in cooler air. So, warm air is pushed into the balloon. This causes the balloon to rise. Then the air cools. The pilot fires up the burner. Hot air replaces the cool air again. And the balloon stays afloat.

History of Hot Air Balloons

In 1783, two brothers sent a duck, a sheep, and a chicken up in a hot air balloon. It was a short flight. It lasted only a few minutes, but it was long enough to begin an exciting new sport. That sport was hot air ballooning.

The brothers used burning straw and manure. This gave the balloon enough heat to rise. Today, hot air balloons use propane gas.

Hot air balloons were once a popular fad!

Kite Science

In the 1700s, two scientists tied thermometers to kites. They wanted to record the air temperature at high altitudes. This was the first time someone used kites to get scientific information.

Wegener enjoyed teaching. He wrote a book based on his exciting lectures about the movement of the earth. Many colleges used his book. He had done many things in his young life that were interesting to read about. He had many good ideas, too. He was only 30 years old!

He didn't stop exploring, either. In 1912, he went back to Greenland for more exploring. He went on the longest crossing of the ice cap ever made by foot. It was a very dangerous trip.

He and three other explorers almost died! A **glacier** they were climbing came loose. They had to spend the entire winter on the ice cap. Wegener used this time to collect information. He learned about storms over the Arctic. He became an expert on Arctic meteorology.

Wegener's travels in Greenland helped him become one of the world's most knowledgable Arctic experts.

Fun Fact

There is an Alfred Wegener Institute in Germany. The institute does polar and marine research. It is carrying on Wegener's work in geology, or Earth science. It conducts research in the Arctic. It gives assistance for polar expeditions.

Marie Tharp [1920–2006]

Marie Tharp studied geology in the 1940s. World War II was going on. Many men were sent to war. Because of this, women had new opportunities. The University of Michigan allowed women to enter the geology department. She graduated with honors. She loved research. She made maps of the ocean floor. The maps were of segments of the ocean. Her partners gathered data from boats. She pieced the maps together. That's when she noticed the mountains did not match up. There was a gap down the center. It had peaks on each side. She thought it might be a **rift valley**. That idea was similar to Wegener's continental drift idea. Her partners thought it was not possible. Soon, though, Tharp's ideas were confirmed. They published the map in 1977. Tharp and her partners received the Hubbard Award in 1978 for their work.

Else Koppen

When Wegener returned from Greenland, he married Else Koppen. She was the daughter of a well-known meteorologist. They had met five years earlier. When he went to Greenland, Koppen moved to Norway. She wanted to be closer to him. Wegener and his wife had three daughters.

Wegener served in World War I. He was injured two times! He was released from combat duty. He then worked with the Army weather-forecasting service.

Later, he took a job that had belonged to his father-in-law. He became the director of a meteorological research department. He did experiments there. He recreated lunar craters. He did this by throwing objects at different substances on the ground. He showed that craters were likely the results of something hitting them. He wrote papers about these experiments. He became quite well known for these papers.

Scientific Theories

A scientific idea is called a theory. Theories take time to develop. They need a great deal of **evidence** before they are accepted or rejected. And even if they are accepted, a theory is always open to debate. More **evidence** might challenge the theory. Theories do not grow up to become facts. Facts are what we observe in nature. Theories describe how the facts work. They are accepted until a better description can be made.

lunar craters

Geologists Rock!

Would you like to learn how to predict earthquakes? Do you want to know how to predict volcanic eruptions? How about explaining why fossils of marine animals can be found atop Mt. Everest? Does working outdoors sound fun? Would you like to see the world? If so, **geology** might be the career for you. **Geologists** study the forces that shape the earth. They study their effects on mankind. They search for oil, water, and gas. They try to predict earthquakes and tsunamis.

Puzzle Pieces

In his studies, Wegener noticed something about the earth. His discovery would one day change the world of science.

Wegener saw that the coastlines of South America and Africa fit together. He saw this while looking at a map. It looked like a puzzle.

He noticed that many of Earth's **continents** looked this way. They seemed like giant jigsaw puzzle pieces that could fit together. Others also noticed this pattern, but they thought it was a **coincidence**. Wegener thought it was something more.

Bears are found on many different continents around the globe.

Wegener thought that all of Earth's continents might have once fit together.

Wegener noticed something else while he was reading a book in the library. The book included lists of similar plant and animal fossils that were found on opposite sides of the ocean. How could this be?

He started looking for more cases like this. He wanted to see if the oceans separated other living things that are similar. If so, the pieces of the puzzle just might be much more than coincidence.

How Many Continents?

In the United States, students are taught that there are seven continents (Africa, Antarctica, Asia, Australia, Europe, North America, and South America). In other parts of the world, students learn that there are only six. Sometimes North and South America are combined and called the Americas. Some schools combine Europe and Asia. They call it Eurasia. There is no good definition of what a continent is. That is why there is so much disagreement.

Wegener thought that fossils found in some areas could not have survived the climate there. Remember, he was a meteorologist. He knew all about climates. He thought some fossils found in Antarctica could never have survived the cold. He figured they must have come from a place closer to the **equator**.

PERMIAN PERIOD
225 MILLION YEARS AGO

Almost 100 years ago, Wegener said all of Earth's continents used to be connected. He called this single landmass **Pangea**. It means "all the earth."

JURASSIC PERIOD
135 MILLION YEARS AGO

He was not the first person to suggest this. But he was the first to try to prove it. He used different sciences. Today, this scientific theory is called plate tectonics. It is widely accepted. Back in Wegener's day, geologists did not believe him. Even his father-in-law did not believe him. He was upset that Wegener had strayed from meteorology. He didn't want Wegener to be working in a new and unknown science.

TRIASSIC PERIOD
200 MILLION YEARS AGO

CRETACEOUS PERIOD
65 MILLION YEARS AGO

PRESENT DAY

Moving the Coal

Miners gather coal in Pennsylvania. How did the coal get there? It was formed from tropical plants that lived near the equator many years ago. The land that is now Pennsylvania was once at the equator!

Proving the Theory

Wegener spent a great deal of time trying to solve the mystery of Pangea.

Wegener needed evidence to prove his theory. He looked at the United States and Europe. He found similar plant and animal fossils in both places. They were on the eastern coast of the United States and the western coast of Europe. He figured that if the same fossils were in both places, then maybe those continents had been together at one time. He thought this happened about 300 million years ago.

Wegener also looked at South America and Africa. He found evidence that the rocks of the eastern coast of South America matched the rocks of western Africa. This was good proof that Pangea had really existed.

So why did other geologists not believe his theory? Wegener did not have a good explanation for how the continents moved. He figured that the continents moved through the ocean floor. He thought it was like ships going through sheets of ice. He had seen ships do this in the icy waters of the Arctic Ocean.

Wegener used fossils like these to try to prove his theory.

Special Delivery

Wegener has appeared on many postage stamps. This stamp is from Austria.

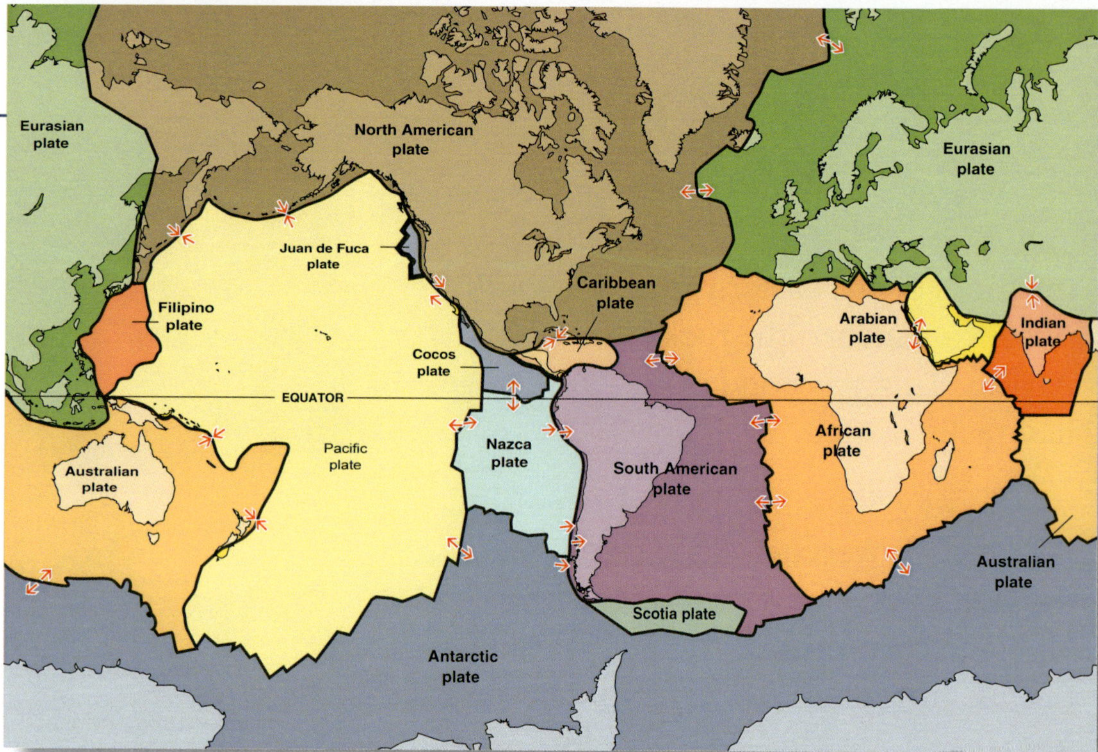

This map shows Earth's tectonic plates and the direction each is moving.

Wegener thought two forces caused the movement. The first was the spinning motion of Earth. The second was the pull on Earth by the moon and the sun's gravity. Most geologists thought that these forces were not strong enough to move the continents. They believed that the ocean floor was too thick and strong. They thought Wegener's theory about how the continents move was wrong. His ideas were rejected.

Wegener made good points about the similar geological features, puzzle-piece shapes, and similar fossils. Many geologists had another explanation for these points. Their explanation was land bridges. It was thought that land bridges connected the continents at one time. These land bridges were now sunk beneath the ocean. There had been some land bridges, but today we know that there weren't enough to explain all of Wegener's data.

Janet Sumner

Today we use plate tectonics, or the movement of Earth's plates, to explain earthquakes and volcanoes. This has opened whole new areas of science to study. For example, today there are scientists who are **volcanologists**. They study volcanoes. Janet Sumner is one of those scientists.

Sumner likes adventure. She enjoys extreme sports. She skydives and goes scuba diving. And she spends a lot of time on active volcanoes. Good thing she's an adventurer!

Sumner is especially interested in volcanic fire-fountain eruptions. **Lava** from these fountains is very dangerous. It is fast flowing and can travel far distances. She has also found that syrup and candle wax behave a lot like **magma**. Using these liquids, she found how clots of magma produce lava.

Wegener didn't know that plate tectonics cause volcanoes. But his work has been a big help to scientists like Sumner.

Some scientists supported Wegener. One geologist thought Wegener's ideas explained the similarities between Africa and South America. Another geologist believed that Wegener's ideas explained his own observations in the Swiss Alps.

Wegener didn't live to see geologists accept his theories. He was finally offered a full professorship to teach meteorology. A few years later, he went to Greenland again. The trip was his fourth, and his last. He went there to study the jet stream. While there, he celebrated his fiftieth birthday. He died after bringing supplies to a scientific station. He was returning to his base camp. The harsh winter weather was most likely to blame.

Wegener's wife, Else, was deeply saddened by his death. She wrote a book in memory of her husband and his work. She lived to be 100 years old. She died in 1992.

Island or Continent?

Greenland is an island, and a very large one at that. Its area is 2,166,086 square kilometers (836,330 square miles). In fact, Greenland is the largest landmass to be called an island rather than a continent.

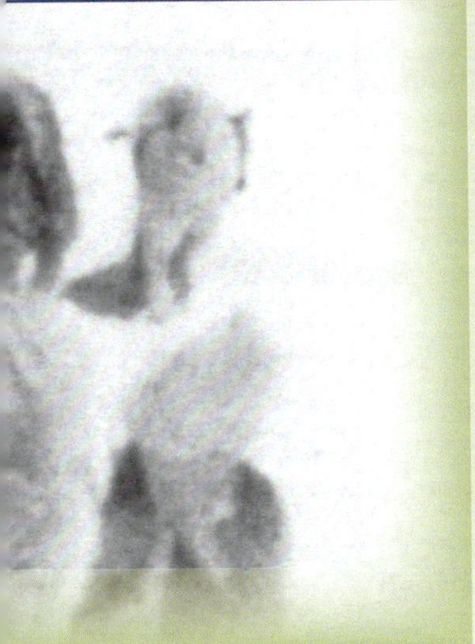

Wegener during a trip to Greenland

Jet Stream

The jet stream is like a river of wind in the atmosphere. It is about 6,000 to 9,000 meters (20,000 to 30,000 feet) above Earth's surface. The jet stream helps move storms around the earth. It also helps to push jet aircraft. In the Northern Hemisphere, the jet stream travels from the west to the east. A plane flying from California to New York will be pushed more quickly to its destination than a plane traveling in the opposite direction. One flying from New York to California will be slowed down by the jet stream.

Jet streams are usually located in the northern regions of the earth. That's why Wegener was studying them in Greenland. During very cold weather, they can travel south. They bring storms across southern areas.

Wegener's Theory Now

Today, Wegener is recognized for many things. He was a record-holding balloonist. He used weather balloons to track air masses. He wrote textbooks that were widely used throughout Europe. He gave the start to modern theories about rainfall. Craters on both Mars and the moon are named for him. There has even been an asteroid named after him!

Wegener is best remembered as an early developer of the theory of plate tectonics. It took years for others to catch on, though. During his lifetime, Wegener didn't have the evidence he needed. He couldn't prove to geologists that Pangea did exist. He couldn't prove that the continents move. The technology was not available.

Since Wegener's time, other scientists have used his work to come closer to a true understanding of plate tectonics

Harry Hess (1906–1969)

Harry Hess taught geology. He helped start the study of plate tectonics. He believed much of what Wegener had proposed. He didn't agree with Wegener about how the earth moved, though.

Hess was in the U.S. Navy during World War II. He was able to do research while at sea. He surveyed the Pacific Ocean while moving from one battle to the next. He outlined how **seafloor spreading** works. He said magma oozes up. It forms a new seafloor. The old and new floors then move away from each other. It is a very slow process.

Unlike Wegener, Hess was able to see his work accepted during his lifetime.

Now we know that Earth is actually made up of 15 plates. The plates are part of Earth's crust. Some of the plates make the continents. Some make the ocean floors. These plates ride around on top of Earth's **mantle**.

Alfred Wegener

It was not until the mid-1900s that scientists finally found enough evidence to show how the continents move. This is the evidence that Wegener needed.

He didn't live long enough to find it for himself. That is the nature of science. The work of one scientist builds upon another. In the future, scientists are sure to learn more and more about the work that Wegener started.

a lava cascade

Geophysicist: Mary Lou Zoback

Western Earthquake Hazards Team (USGS)

Zoback's favorite subject in school was math.

Shake, Rattle, and Roll

Mary Lou Zoback is really into her job. She even takes work to the movies. Sometimes she sees mistakes in movies. They are called bloopers. In one movie she noticed a big blooper. The mountains were in the wrong place!

Zoback became interested in geophysics in college. During that time, many discoveries were being made. Scientists were learning more about how Earth's surface moves. This movement causes earthquakes.

Zoback's goal is to predict earthquakes. How? She starts by using instruments that measure how much the ground shakes during an earthquake. And she uses information from satellites. The satellites can measure Earth's surface from space.

Her favorite parts of her work are the surprises. "Often you look for one thing and something else completely different will jump out at you," she tells Sally Ride Science. "That's the fun part of science."

Seismometers such as this can measure waves of energy.

Being There

If you were a geophysicist, you would use physics to study Earth and how it changes. You might . . .

- find underground water.
- climb volcanoes.
- investigate where mountains came from.

Experts Tell Us . . .

"If you get first-hand experience with science, I think you'd get hooked."

Is It 4 U?

How do you like surprises? Do you like it when you're looking for one thing and find something different?

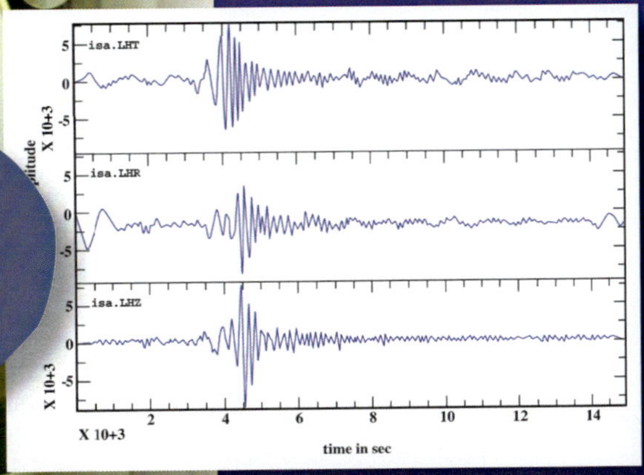

Did You Know?

Every day there are thousands of small earthquakes around the world.

Lab: An Eggsample Look

Geologists dig deep to get an idea of what Earth looks like inside. Unfortunately, they cannot dig all the way to the core. However, geologists can use earthquakes to help them "see" inside Earth. When an earthquake happens, it sends shock waves through the planet. Earthquake waves travel differently when they pass through liquids and solids. A geologist can tell if it was solid or liquid material the waves passed through when they receive earthquake data. By recording a number of earthquakes, geologists have a pretty good idea of what our Earth looks like inside. A core sample would be more accurate, but is not possible today. To get an idea of a core sample, try this experiment. You will drill core samples from an egg to see what it looks like inside.

Materials

- hard-boiled egg
- clear plastic drinking straw
- plastic knife
- scissors

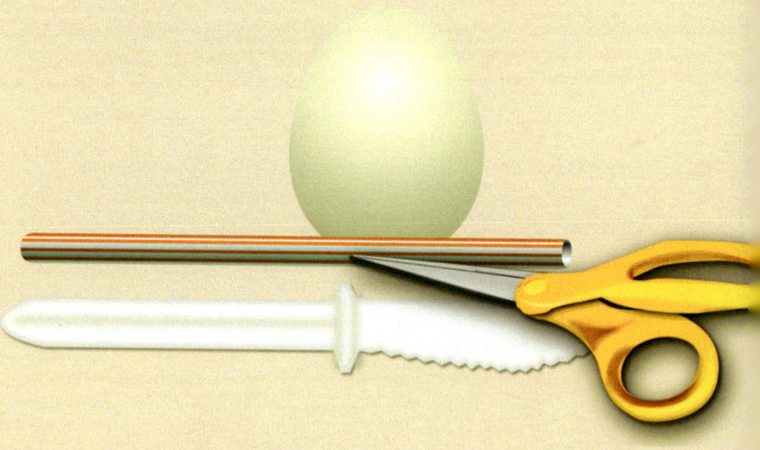

Procedure

1 Crack and peel the shell off of the hard-boiled egg.

2 Hold the egg in one hand and insert the straw into the top of the egg with your other hand. Slowly but firmly, press the straw through the center of the egg and out the other side.

3 When the straw exits the other side of the egg, you will see parts of the egg in the straw. This is your core sample. As you continue to push, you will see a part of the straw that doesn't have any egg in it. Cut the straw at this point.

4 Pull the remaining part of the straw out of your egg. You can dig another core sample from a different location on your egg with the rest of the straw. Try entering the egg from a different angle.

5 Again, when the straw exits the egg, cut it off when you no longer see any core sample inside of it.

6 Use your scissors to cut open your straw pieces. Examine your core samples.

7 Draw a picture of what you think your egg looks like inside, based on your core samples.

8 When you have drawn your picture, slice the egg open with the plastic knife to see how close your drawing is to the actual egg.

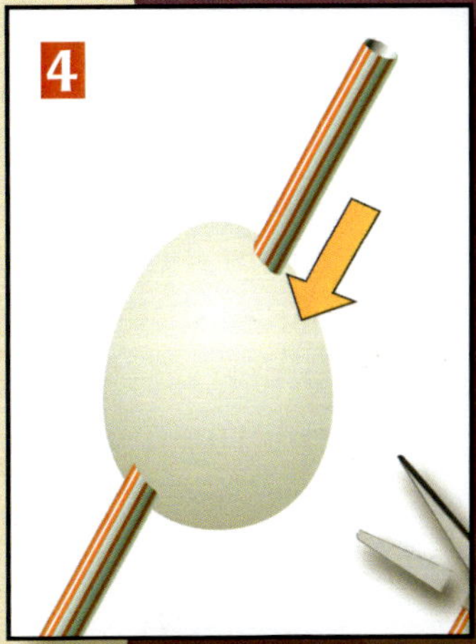

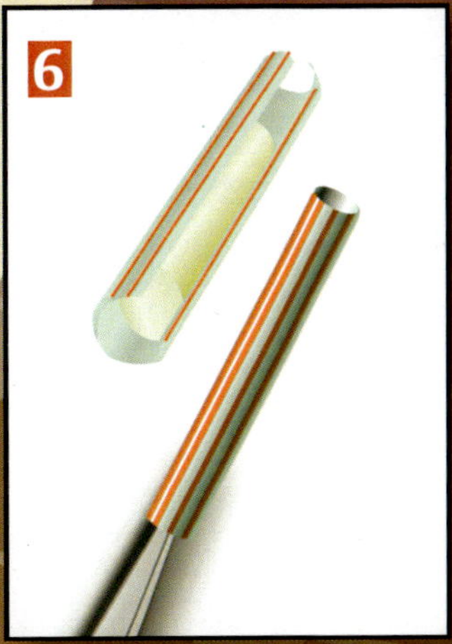

Glossary

astronomy—the scientific study of the universe beyond Earth

coincidence—remarkable series of events or occurrences which seem to have no connection

continent—a single large area of land

continental drift—the theory that continents shift positions on the earth's surface, moving across the surface rather than being fixed, stationary land masses

Earth science—the scientific study of the structure, age, etc. of the earth

equator—an imaginary line drawn around the middle of the earth at an equal distance from the two poles

evidence—an available body of facts or information that supports a theory

expedition—an organized journey for a particular purpose

geologist—a scientist who studies the earth's structure, substance, and history

geology—the study of rocks and other substances that make up the earth's surface and interior

glacier—large mass of ice and snow that slowly moves over land

lava—molten rock that reaches the earth's surface

magma—molten rock in the earth's crust

mantle—layer of earth below the crust

meteorology—the scientific study of processes in the earth's atmosphere that cause weather conditions

Pangea—single landmass made of all of today's continents; means "all the earth"

plate tectonics—the theory that the earth's crust is made of rigid plates that move about on the earth's surface

rift valley—a valley that has been formed along a rift

seafloor spreading—the process by which new oceanic crust is formed by the convective upwelling of magma at mid-ocean ridges

theory—a system of ideas intended to explain observations

volcanologist—a scientist who studies volcanoes

Index

Africa, 12–13, 16–17, 20
Alfred Wegener Institute, 8
America, 13
Antarctica, 14–15
Arctic, 8–9
Arctic meteorology, 8
Arctic Ocean, 17
Asia, 13
asteroid, 22
astronomy, 4
Austria, 17
Berlin, 4
California, 21
continental drift, 6, 8–9, 12–25
continents, 12–20, 22–25
Earth science, 8
earthquakes, 11, 19, 26–28
equator, 14–15
Europe, 13, 16, 22
evidence, 10, 16–17, 22–25
expedition, 4–5, 8
fossils, 11, 14–18
geologist, 11, 14–23, 28
geology, 8–11, 22–23
Germany, 4, 6–9
glacier, 8
Greenland, 4–6, 8, 10–11, 20–21
Hess, Harry, 23
hot air balloon, 5–7
jet stream, 20–21
Koppen, Else, 10, 20
land bridges, 18
lava, 19
magma, 19, 23
Mars, 22
meteorology, 5–8, 10, 14, 20–21
molten magma, 24–25
moon, 18, 22
Mt. Everest, 11
Navy, 23
New York, 21
North America, 13
Pacific Ocean, 23
Pangea, 14–18, 22
Pennsylvania, 15
plate tectonics, 6, 14–19, 22–25
seafloor spreading, 23
South America, 12–13, 20
stamps, 17
Sumner, Janet, 19
Swiss Alps, 20
Tharp, Marie, 9
theory, 6, 10
United States, 13, 16
University of Michigan, 9
volcanoes, 19, 26–27
volcanologist, 19
Wegener, Anna, 4
Wegener, Kurt, 6
Wegener, Richard, 4
Zoback, Mary Lou, 26–27

Sally Ride Science

Sally Ride Science™ is an innovative content company dedicated to fueling young people's interests in science. Our publications and programs provide opportunities for students and teachers to explore the captivating world of science—from astrobiology to zoology. We bring science to life and show young people that science is creative, collaborative, fascinating, and fun.

Image Credits

Cover: The Granger Collection, New York; p.3 Alfred Wegener Institute; p.4 (top) Alfred Wegener Institute; p.4 (left) Imagno/Getty Images; p.4–5 Tyler Olson/Shutterstock; p.5 (left) Alfred Wegener Institute; p.5 (right) Paul Laragy/Shutterstock; p.6 (top) Ralph Biggör/Shutterstock; p.6–7 Jean Becker/Sygma/CORBIS; p.7 (top) Topical Press Agency/Getty Images; p.7 (bottom) Library of Congress; p.8 (left) Mary Evans Picture Library/Alamy; p.8–9 Bildarchiv Monheim GmbH/Alamy; p.9 (left) Bruce Gilbert/Columbia University; p.9 (right) Public Domain; p.10 (top) Rick Reason; p.10 (bottom) Diego Barucco/Shutterstock; p.11 Jason Bodine; p.11 (background) Kavram/Shutterstock; p.12 (top to bottom) Regien Paassen/Shutterstock; Shootov Igor/Shutterstock; Jens Stolt/Shutterstock; Susan Kehoe/Shutterstock; Shona Charlton/Shutterstock; p.12–13 Tim Bradley; p.13 Wojciech PLONKA/Shutterstock; p.14–15 Tim Bradley; p.15 (top) Graham S. Klotz/Shutterstock; p.15 (bottom) Andre Nantel/Shutterstock; p.16 (top) USGS; p.16 (bottom) The Granger Collection, New York; p.17 (left) The Granger Collection, New York; p.17 (top to bottom) Alexey Kryuchkov/Shutterstock; Paul Maguire/Shutterstock; Falk Kienas/Shutterstock; p.18 Tim Bradley; p.19 Courtesy of Janet Summer; p.19 (background) Hoa-Qui/Photo Researchers, Inc; p.20–21 The Granger Collection, New York; p.21 Tim Bradley; p.22 (top) Michael Schofield/Shutterstock; p.22 (left) SPL/Photo Researchers, Inc.; p.23 Princeton University Archives; p.23 (background) Michael Schofield/Shutterstock; p.24–25 Ashok Rodrigues/iStockphoto; p.26 Courtesy of Megan Lacey Zoback; p.27 (left) Lockhaven University of Pennsylvania; p.27 (right) Tim Bradley; p.28 (top) J.D. Griggs, Hawaiian Volcano Observatory, U.S. Geological Survey; p.28–29 Nicoll Rager Fuller